Functions in Mathematics

Introductory Explorations for Secondary School Teachers

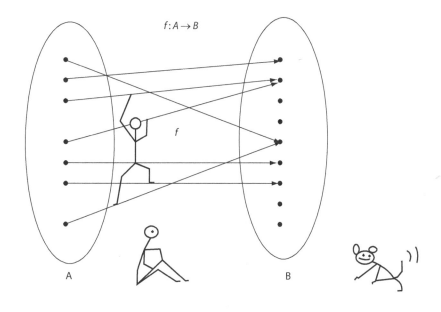

By Mark Daniels and Efraim R. Armendariz

University of Texas — Austin

cognella™
San Diego, CA

Bassim Hamadeh, Publisher
Michael Simpson, Vice President of Acquisitions
Christopher Foster, Vice President of Marketing
Jessica Knott, Managing Editor
Stephen Milano, Creative Director
Kevin Fahey, Cognella Marketing Program Manager
Rose Tawy, Acquisitions Editor
Jamie Giganti, Project Editor
Erin Escobar, Licensing Associate

First published in the United States of America in 2012 by University Readers, Inc.

Trademark Notice: Product or corporate names may be trademarks or registered trademarks, and are used only for identification and explanation without intent to infringe.

16 15 14 13 12   1 2 3 4 5

Printed in the United States of America

ISBN: 978-1-60927-168-8 (pbk) / 978-1-60927-326-2 (br)

www.cognella.com   800.200.3908

Contents

UNIT THREE: EXPLORING FUNCTIONS IN OTHER SYSTEMS   61

Acknowledgment

The authors would like to thank Mr. Harry Lucas, Jr. and the Educational Advancement Foundation for their support of our projects and for their promotion of inquiry-based teaching and learning in mathematics at all grade and university levels.

Preface

Functions in mathematics: introductory explorations for secondary school teachers

About this Text

The presentation of topics in this text is done in a "discovery-based" format so as to invite the learner/reader to actively participate in the Explorations that develop the material in a logical manner.

This manuscript is intended to be used primarily for a one-semester lower-division mathematics course. The course assumes that students have a working knowledge of the basic concepts of Calculus associated with differentiation and integration, but this is not a rigid prerequisite. Due to the collaborative nature of the delivery of the course, we have had students succeed in the course who have not had a Calculus background. The topics covered in this text serve to synthesize important topics from secondary mathematics curriculum and to connect these topics to university-level mathematics. We believe that the material and mathematical connections made in this text are worthwhile and extremely well suited for a discovery-based mathematics course for preservice secondary teachers. The course topics were born out of notes used for an introductory mathematics course that is taught to lower-division mathematics majors in the UTeach Program in the College of Natural Sciences at the University of Texas. UTeach mathematics majors, in addition to earning an undergraduate degree in mathematics, are seeking state teaching certification in middle school or high school mathematics.

Suggestions for Using this Text

Instructors of a course using this text are encouraged to present the material of the book in an inquiry-based method. That is, allow students to work through the Explorations of the text collaboratively and devote considerable class time to student presentation of results. The instructor is further encouraged to take on the role of facilitator in the course in that one of the goals of the course would be to bring students to the point where students would not simply present results, but also challenge and correct each other concerning the formal logic, mathematical language, and methods employed in their presentations. This is a book and a course about getting students to think deeply and logically about fundamental ideas in mathematics.

Further, instructors are encouraged to add any relevant topics, exercises, and explorations to those presented in the text that are felt necessary based on student make-up and individual goals for the course. The authors welcome any suggestions along those lines. Please feel free to use the material of the text as a springboard for exploration of tangential topics and connections based upon student interest, deficiencies, and discussion.

Lastly, to the student, we encourage students to view the topics and Explorations of the text as vehicles to be used toward thinking deeply about concepts and connections between concepts that you may have seen before but not in the same depth or context.

Introduction

The layout of this text is presented in sections labeled by "Lesson" of instruction. It is assumed that these Lessons correspond to class periods of approximately one and one-half hours in duration. Even so, it may be the case that you don't use all of the Explorations within, or that you don't finish the Explorations of each Lesson, or that you are working with a shorter class period in which some Explorations will continue over multiple days. The Lesson sections are numbered in order and do not include such things as time allotted for class tests or discussion of tangential topics.

The key to experiencing this course is to approach the associated teaching and learning with a flexible mindset. Let the Explorations take you in many directions based upon the presentation and discussion of the material under investigation. Consider and be open to the fact that there are often multiple ways to approach or obtain the desired result for a given Exploration. Much of the learning in this course will come from listening to others' justifications and explanations of how a result was obtained.

Let us begin!

UNIT ONE
FUNCTIONS, RATES, AND PATTERNS

Lesson 1
Getting Started

The three Explorations of this section are meant to "get you started" on the right track regarding the expectations of this text. This text is meant to accompany a mathematics course that, among other things, is about "thinking." By this we mean it is our intention that the activities presented within will entice you to think deeply about some of the mathematics you've encountered previously, about new ideas presented, and about the connections between the two. With this in mind, let's get started.

Exploration 1.1: A Common Cube Conundrum

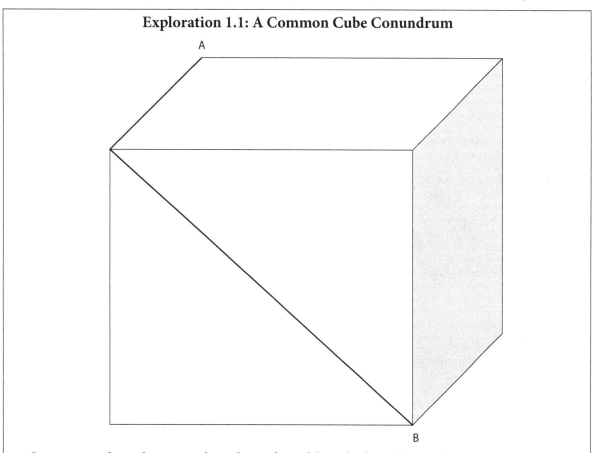

Suppose one desired to move along the surface of the cube from the top far corner (point A) to the near bottom corner (point B). Would the dark line path marked in the picture be the shortest route from point A to point B? If you agree, justify your answer. If you do not agree, describe the shortest path and justify your answer. Is the shortest path solution that you have chosen unique?

Exploration 1.2: The Efficient Waterline

The city wishes to connect two houses to an existing water supply line using a single pump at the supply line and a minimal amount of pipe extended to each house. The houses are located at different distances from the street (as pictured). Where should the pump connection be located? Of course, you must justify your answer.

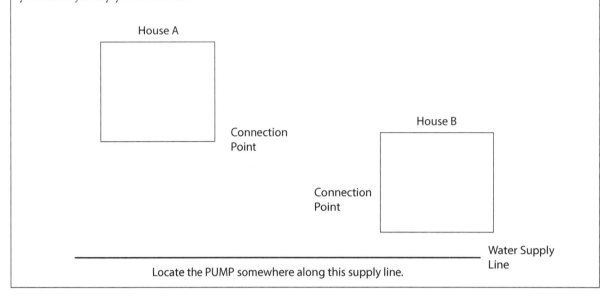

As mentioned previously, it is important that you also make connections between mathematics concepts you are asked to remember and that you newly learn in this course. The next Exploration is a good example of making connections between topics and concepts in mathematics.

Exploration 1.3: Making Connections

Given the three topics listed below, devise a visual, verbal, and algebraic way of connecting the concepts:

1. The distance formula
2. The standard equation of a circle (not centered at the origin)
3. The Pythagorean Theorem

Can you think of further related mathematics topics that can be extended from these three?

Lesson 2

What is a Function?

In this section, we will take a deeper look at the concept of *function* in terms of a function's definition, type, and general properties. You have encountered and worked with functions in your mathematical experiences so far, but have you really thought much about what constitutes a *function*?

Exploration 2.1: Function?

Work in groups to answer the question, "What is your definition of a *function*?" Each group should agree upon and present one definition.

Each group should also display their agreed-upon definition for the class and all definitions should be recorded by each student. At this point, however, a formal definition of function should not yet be agreed upon and formulated.

Exploration 2.2: Function an Identification Activity 1

Function-Identify?

Decide, in groups, which of the relations listed below are examples of functions; justify your answers.

	Function?		Justification
1	$y = -x^2$		
2	**x**	**y**	
	2	4	
	-6	-12	
	13	26	
	-57	-114	
3	(5, 6) (3, 2) (5, 1)		

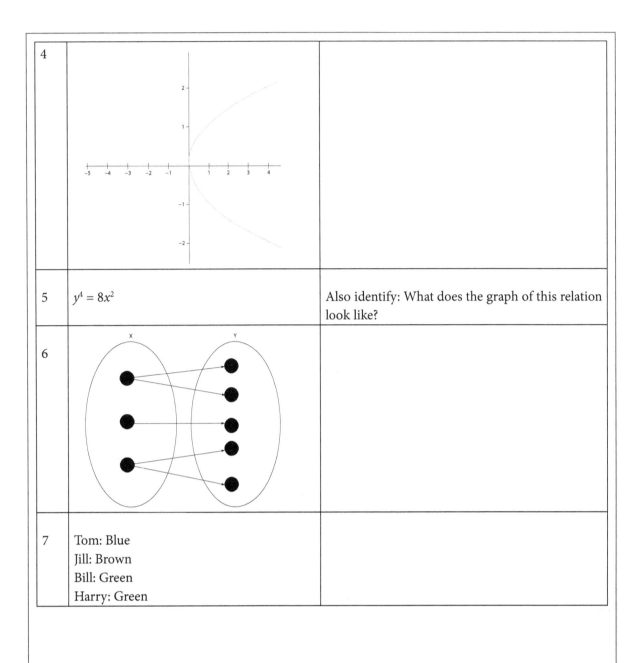

4		
5	$y^4 = 8x^2$	Also identify: What does the graph of this relation look like?
6		
7	Tom: Blue Jill: Brown Bill: Green Harry: Green	

Exploration 2.4: "Mapping" a Map—Activity 1

Find a map of your choice. This could be a campus map, a local street map, or state map, for example. Use any the information supplied in the map or any subset of the information provided to create a function. Be sure to explain your reasoning behind the creation of your function.

Exploration 2.4: Function Identification—Activity 2

As you discuss these problems, consider the meaning of the symbols in the set-builder notation. These symbols should be discussed in class as part of the next section.

1. Let $A = \{a, b, c\}$, $B = \{4, 5, 6\}$, and $f = \{(a, 6), (b, 4), (c, 6)\}$.   Is f a function from A to B?

2. Let $A = \{1, 2, 3\}$, $B = \{c, d, e\}$, and $g = \{(1, d), (2, c), (1, e)\}$.   Is g a function from A to B?

3. Let M be the set of all museums, N the set of all countries, and $L = \{(m, n) \in M \times N \mid \textit{the museum m is in the country n}\}$.

   Is L a function from M to N?

4. Let D be the set of all dogs, and let $C = \{(d, o) \in D \times D \mid \textit{the dog d is a parent of the offspring o}\}$.

   Is C a function from D to D?

TAKE-HOME EXERCISE: Reflect on your informal work in trying to describe what a function is and consider the various group definitions of *function* presented. Now revise the definition you originally created for describing a function in order to develop a more refined definition. Explain your reasons for refining (or not refining) your function definition.

Lesson 3

Functions and Types of Functions

We will now attempt to formalize our definition of function by providing three textbook definitions of the concept of function. Note the use of the symbols in each definition. As you read over the three provided definitions of a function, you are asked to consider the commonality and differences between these definitions and the one that you have previously written. Three examples of how a function might be defined are

Given two sets A and B, the set $A \times B$ consists of all ordered pairs (a, b) where $a \in$ A, $b \in$ B. A subset of $A \times B$ is called a relation. Thus:

1. A function from **A** to **B** is a pairing of elements in **A** with elements in **B** in such a way that each element in **A** is paired with exactly one element in **B**.
2. A function f from **A** to **B** is a rule or relation between **A** and **B** that assigns each element $a \in$ **A** to a unique element $b \in$ **B**.
3. A function f from **A** to **B** is a subset of the Cartesian product $A \times B = \{(a, b) \mid a \in$ **A**, $b \in$ **B**$\}$ such that b is unique for each $a \in$ **A**$\}$.

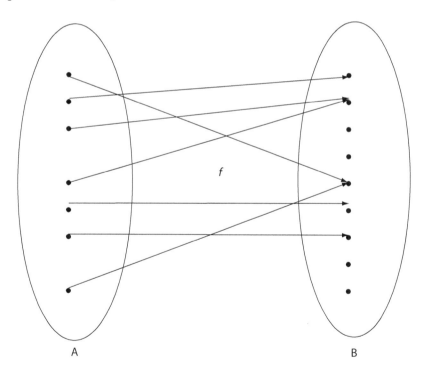

4. A function between **A** and **B** is a nonempty relation $f \subseteq A \times B$ *such that if* $(a, b) \in f$ *and* $(a, b') \in f$, *then* $b = b'$. The domain of f is the set of all first elements of members of f and the range of f is the set of all second elements of members of f. Symbolically:

$$domain \ f = \left\{ a : \exists b \in B \ni (a, b) \in f \right\}$$

$$range \ f = \left\{ b : \exists a \in A \ni (a, b) \in f \right\}$$

The set B is referred to as the codomain of f. If it happens that the domain of f is equal to all of A, then we say f is a function from A into B and we write $f : A \to B$.

EXERCISE: Consider why it is or is it not important to have a precise definition of the term "function."

Historical Notes

1. Although the notion of a *function* dates back to the seventeenth century, a relation-based definition as we use today was not formulated until the beginning of the twentieth century. The concept of mathematical relations first appears in the text *Geometry*, written by Rene Descartes in 1637, and the term "function" was introduced about fifty years later by Gottfried Wilhelm Leibniz. It was Leonhard Euler, in the eighteenth century, who first used today's notation $y = f(x)$. Finally, it was Hardy who, in 1908, defined a function as a relation between two variables x and y such that "to some values of x at any rate correspond values of y."

2. French author Nicolas Bourbáki is not a single author but a group of authors who came together in the late 1950s in an effort to standardize the language of modern mathematics. That language is used in the definitions that follow. Among the group was mathematician John Tate, who later became a faculty member at the University of Texas at Austin.

In addition to knowing whether one is working with a function or not, it is often useful to know the type of function under investigation based upon the mapping properties of the function.

Given two sets A and B.

DEFINITION A function $f : A \to B$ is called **surjective** (or is said to map A onto B) if $B = $ range f.

DEFINITION A function $f : A \to B$ is called **injective** (or **one-to-one**) if, for all a and a' in A,

$f(a) = f(a')$ implies that $a = a'$.

DEFINITION A function $f : A \to B$ is called **bijective** if it is both surjective and injective.

If a function is both surjective and injective, then it is said to be particularly "well behaved."

Exploration 3.1: Types of Functions

1. Devise and explain two examples for each of a surjective function, an injective function, and a bijective function.
2. Characterize the function f pictured here:

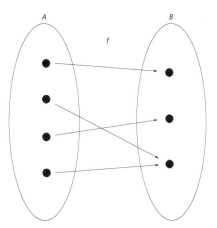

The Ubiquitous Quadratic Function

The quadratic function

$$f(x) = ax^2 + bx + c$$

can certainly be said to be present everywhere in secondary mathematics curriculum. It is likely to be the function that you will investigate the most with your students as a secondary mathematics teacher of various levels of mathematics. Due to this fact, we will explore some of interesting properties of this function throughout this text.

In the next Exploration, we will try to devise a way to visualize the location of the roots of a quadratic when it is the case that the roots are complex. In order to accomplish this task, we will allow the domain of the function to be the complex numbers. It is assumed that you are comfortable with working with complex numbers and performing some basic field operations in this system.

Exploration 3.2: Complex Roots Visualization

Consider the graph of $f(x) = x^2 + 4x + 7$

1. Graph $f(x)$ in detail in the xy plane of the coordinate system provided on the next page without using a graphing calculator.
2. What conclusion can you make about the roots of $f(x)$?

3. Suppose that we can use the complex numbers as the domain for $f(x)$.

    a. Show that $f(-2 + \sqrt{3}\,i)$ is a root for f.

    b. Show that $f(-2 + 5i)$ is a real number and that $f(3 + 5i)$ is *not* a real number.

    c. Try a few more complex values and make a conjecture about values of a and b for which $f(a + bi)$ is a real number. Explain how you arrived at your conjecture and prove that it is true.

d. Lastly, can you visualize and draw a graphical representation of what your above answers imply about the real-valued outputs of *f* with regard to the inclusion of a complex domain? Try to do this using the 3-dimensional coordinate system provided below.

Axes template: (be sure to scale axes):

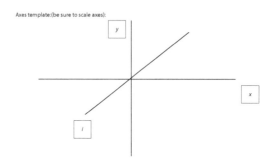

Axes template:(be sure to scale axes):

CONNECTIONS EXERCISE:

Based on your previous experience and the exploration that you completed involving "roots" and complex numbers, state four ways one might find the roots (or zeros) of a quadratic function.

Now, using the general formula $ax^2 + bx + c = 0$, prove that one method that you have mentioned

is really a general case of the other (i.e., derive the quadratic formula).

Lesson 4

A Qualitative Look at Rates

By considering specific velocity-time functions of the motion of a particle as rate of change functions and then trying to draw 1) a rate of change function of the rate of change function and 2) a function that "undoes" the given rate of change function. Of course, we are talking about creating the acceleration-time function and the position-time functions, respectively, when supplied with a velocity-time function. It is assumed that you have some conceptual knowledge of the derivative as a rate of change function and the definite integral as an accumulation function. As you work through Exploration 4.1, try to reconcile both the physical aspects and the mathematics theory behind your answers.

Exploration 4.1: A Qualitative Look at Rates

Work with gaining a qualitative understanding of graphs of functions. In particular focus on the relationships between acceleration, velocity, and position vs. time in terms of rates of change of the functions given.

For each of the following **velocity vs. time graphs** try to sketch what you think the corresponding *position-time* graph will look like. Then try to sketch the corresponding *acceleration-time* graph. Be prepared to present and justify your results.

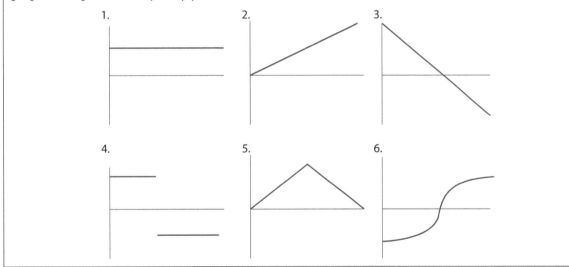

Lesson 5

A Further Investigation of Rate of Change

We will investigate three *separable* differential equations and their solutions. These equations are each used widely to model growth and decay "real-life" situations usually related to populations. Note that these differential equations are simply rate-of-change functions.

Exploration 5.1: A Further Investigation of Rate of Change: Growth and Decay Models from a Differential Equations Point of View

As you work through and present this Exploration, try to focus on being able to present your findings with lucid explanations of the mathematics involved and on using correct mathematics terminology throughout.

Let L, k, and y_0 be constants. We wish to consider the following three *differential* equations,

where y is a function of t.

1. $\dfrac{dy}{dt} = ky$

2. $\dfrac{dy}{dt} = k(y - y_0)$

3. $\dfrac{dy}{dt} = ky(L - y)$

A. Verify that the three functions $y = Ce^{kt}$, $y = y_0 + Ce^{kt}$, and $y = \dfrac{L}{1 + Ce^{-kLt}}$ are solutions,

respectively, to the equations 1, 2, and 3, where C is a constant.

B. Use what you have learned in Calculus to solve each of the three given differential (or rate-of-change) equations for y in order to obtain the given general solution. (The method of Partial Fractions is helpful in solving Equation 3.)

Note: Each Equation is a mathematical model for describing a physical process. Equation 1 represents Simple Growth and Decay, Equation 2 is known as Newton's Law of Cooling, while Equation 3 is a general Logistic Model.

The *Logistic Model* is quite important in population modeling and has application in other branches of mathematics, such as Chaos and Dynamics. As a population model, the constant L is called the *carrying capacity* of the model, and the line $y = L$ is a horizontal asymptote for the solution.

C. Problems

4. The rate of change of the number of wolves $W(t)$ in a population is proportional to the quantity $1500 - W(t)$, where $t \geq 0$ is *time* measured in years and k is the constant of proportionality. Assume, as an initial condition, that when $t = 0$ the wolf population is 500.

    a. Write the differential equation that models this situation and show all steps in solving the equation for $W(t)$.

    b. Find $W(t)$ in terms of t and k.

    c. Use the fact that $W(4) = 800$, to find k.

    d. Find $\lim\limits_{t \to \infty} W(t)$.

5. The rate of growth of a population $B(t)$ of a certain strand of bacterium is proportional to the product of $B(t)$ and the quantity $1200 - B(t)$, where $t \geq 0$ is *time* measured in hours and k is the constant of proportionality.

    a. Write the differential equation that models this situation and show all steps in solving the equation for $B(t)$.

    b. Find $B(t)$ in terms of t and k if $B(0) = 200$

    c. Use the fact that $B(5) = 700$ to find k.

    d. Describe the long-term trend of this population.

Lesson 6
Conic Sections

We will now turn our attention to the *conic sections*. A conic section can be defined as the curve of intersection of a plane with a right circular cone. The curves created by these intersections are the circle, the ellipse, the hyperbola, and the parabola. There are also *degenerate* cases of these curves created, such as the point, the lines, or two intersecting lines. Associated with each of these conics is an analytic definition. As was mentioned previously, the curve generated by the quadratic equation is a parabola. One might ask, for example, how it is that the conic section, the analytic definition, and quadratic function describing a parabola are all related. This is the topic to be explored in this section.

First we will look at the Conic Sections and then their associated analytic definitions. We visualize each conic section as the intersection of a plane with a right circular cone (see Figures 1, 2, and 3).

Figure 1. Parabola.

Figure 2. Ellipse.

Figure 3. Hyperbola.

Depending on the orientation of the plane, the resulting curve can be a parabola, ellipse, circle, or hyperbola (with degenerate forms: point and line or lines). The reader is asked to describe the situation in which a *Circle* arises.

The analytic definitions of the Conic Sections are

DEFINITION: Parabola. We define a parabola (see Figure A) to be the set of all points in the plane that are equidistant from a fixed point and a fixed line. The fixed point is called the focus of the parabola and the fixed line is called the directrix of the parabola. A special point on the parabola is the vertex, the midpoint of the perpendicular segment from the focus to the directrix.

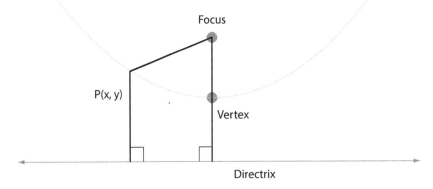

Figure A.

DEFINITION: Ellipse. We define an ellipse to be the set of all points in the plane for which the sum of the distances to two fixed points is constant. The fixed points are called the foci of the ellipse. This definition is illustrated in Figure B. We define the center of an ellipse to be the midpoint of the line segment connecting the foci.

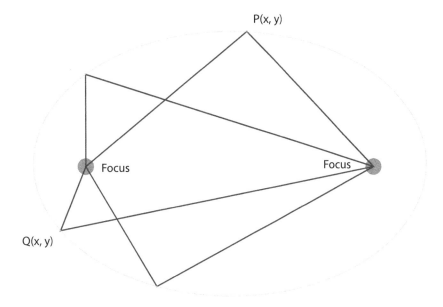

Figure B.

DEFINITION: Hyperbola. We define a hyperbola to be the set of all points in the plane such that the difference of the distances between two fixed points is a constant. This definition is illustrated in Figure C.

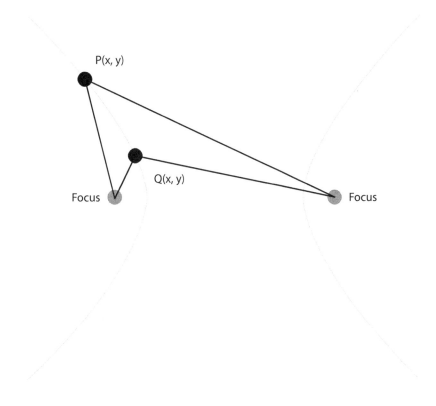

Figure C.

EXERCISE: Based on your discussions in Exploration 1.3, can you provide an analytic definition for the CIRCLE?

Exploration 6.1: The Conics—Equations from Definitions

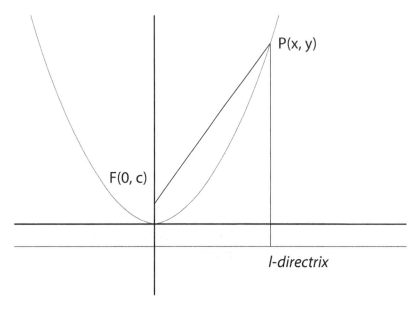

Figure 1.

A. Parabola

Given: A fixed point F and a fixed line l.

The parabola consists of all points P such that $\overline{FP}$ is equal to $\overline{PL}$. (The distance from F to P is the same as the distance from P to the line l; in symbols $\overline{FP} = \overline{PL}$.)

1. Without loss of generality, orient the parabola so that the point F (called the *focus of the parabola*) is on the y-axis and the line l (called the *directrix of the parabola*) is parallel to the x-axis.
2. From Figure 1, let P be a general point on the parabola; what are its coordinates?
3. Use the distance formula to find $\overline{FP}$:

4. Use the distance formula to find $\overline{PL}$:

5. Substitute the expressions in 4 & 5 in the equation $\overline{FP} = \overline{PL}$, and simplify using algebraic techniques in order to derive a general equation that models the parabola.

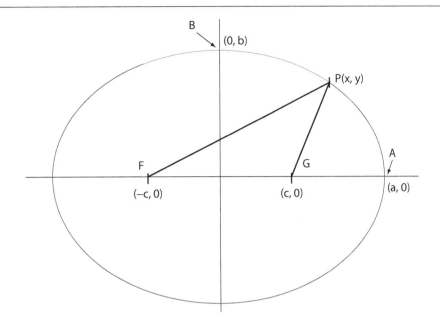

Figure 2.

B: Ellipse

Given: Two fixed points F, G and *a fixed positive number k*. The ellipse consists of all points P such that $\overline{FP} + \overline{GP} = k$.

    The fixed points F and G have coordinates $(-c, 0)$ and $(c, 0)$, respectively. The points A, B are points where the ellipse intersects the positive x-axis and positive y-axis, respectively.

1. Use the distance formula to express the relationship $\overline{FP} + \overline{GP} = k$ in terms of the coordinates of

   an arbitrary point $P(x, y)$, as pictured, which lies on the ellipse. (Your expression should involve a sum of two square roots.)

2. Use algebraic techniques to eliminate the square roots that occur in 1. (Note: This involves squaring twice; it helps to simplify the result obtained by squaring the first time before squaring a second time.) Your expression should involve x, y, c, and k.

3. The next step involves expressing k and c in terms of a and b. Using the fact that A and B are points on the ellipse, verify that $k = 2a$ and $c^2 = a^2 - b^2$.

4. Substitute the values for k and c into your derived equation to obtain the standard equation of the ellipse centered at the origin with semi-major axis of length a and semi-minor axis of length b:

$$\frac{x^2}{a^2} + \frac{y^2}{b^2} = 1.$$

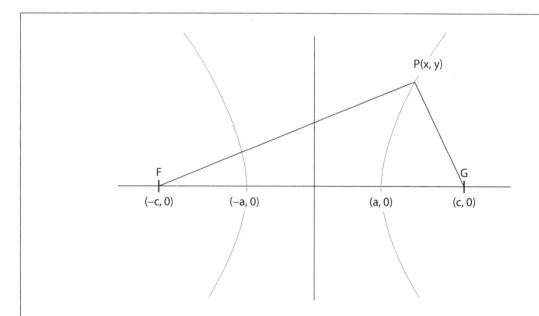

Figure 3.

Hyperbola:

Given: Two fixed points F, G and *a fixed positive number k*. The hyperbola consists of all points P such that $|\overline{FP} - \overline{GP}| = k$.

Follow the procedures outlined for the **Ellipse** noting that the substitution $b^2 = c^2 - a^2$

will lead to the standard form of the hyperbola centered at the origin.

Lesson 7

Spring-Mass Motion Lab

The lab activity that follows enables one to use knowledge of sinusoidal equations from trigonometry to model data and to explore further the concept of "rate of change of a function." Various combinations of commercially available probes and computers or calculators can be used to do the following lab activity. The authors used a Texas Instruments TI-84 graphing calculator connected to a Calculator-Based Laboratory (CBL) unit and a Vernier motion detector. The HOOK program was used on the TI calculator to store and display data. This program can be found on the TI Website.

Lab 1: Spring-Mass Motion Lab

Purpose

You will find a model to represent a "real world" spring-mass system's motion (ignoring damping).

Set-Up and Procedure

Apparatus set-up is pictured below:

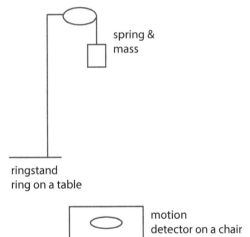

1. Make sure the calculator and the CBL are turned on and that the mass is positioned directly above the motion detector. (For best results, one can tape a small square of paper to the bottom of the mass, so the detector can better "see" the mass.)
2. Start the HOOK (TI©) program on the calculator.
3. Carefully pull the mass down and release it to allow it to oscillate. As soon as you are confident that the motion is smooth and only vertical in direction, press [Trigger] on the CBL (TI Calculator-Based Lab).
4. The motion of the spring-mass will be sent to your calculator as a graph of position vs. time.
5. If you do not obtain a fairly smooth and consistent graph, repeat steps 1 through 4 until you do.

Analysis

You will determine an equation $y = C + A \cos B(x - D)$ that represents the position of the spring

(y) as measured over time (x) and the 1st and 2nd derivatives of this function.

1. Find C in the above equation by using [trace] on your calculator to record the <u>y-value</u> of the first max. and the first min. you encounter on your graph.

    Record C below.

$$C = \frac{y_{max} + y_{min}}{2}$$

    What is the name given to C?

2. A can be found by using the [trace] feature in your calculator just as you did in 1) above.

$$A = \frac{y_{max} - y_{min}}{2}$$

    What is the name given to A?

3. The observed period of the spring's motion can be found by using [trace] to find the <u>x-values</u> that correspond to the first two max.'s on your graph and then computing $x_2 - x_1$. Once you

    have found the observed period for the graph, B can be filled in for the sinusoidal equation above by using:

$$observed\ period = \frac{2\pi}{|B|} \quad \therefore \quad B = \frac{2\pi}{obs.\,per.}$$

    Find B for your equation.

4. D can be found by using [trace] in your calculator. Record the <u>x-value</u> at the first max. on your graph. Record your D value here.

    What is the name given to the D value?

5. Now fill in the equation $y = C + A \cos B(x - D)$ completely for your particular graph and

    write it here.

6. Graph the above equation on your calculator along with the data that you have collected. Do the graphs match? ***At this point, you must show the graphs to your instructor before you can continue.*** Draw a picture of your graph, labeling the period, max, mins, etc.

7. What would y' represent in terms of the motion of the mass-spring? ALSO, find y'. (SHOW ALL WORK)

8. What would y'' represent in terms of the motion of the mass-spring? Find y''. (SHOW ALL WORK).

9. As a check, use NDERIV (in your calculator) to graph the derivative of your original equation. Then graph the y' equation you found above in the same window. Are there any differences? If so why might the two graphs differ?

10. Compare the graphs of y and y' in your calculator. Does their relationship *make sense* based on what you know? Explain.

11. Find the position, velocity, and acceleration of the mass-spring at x = 0.5 sec. (SHOW WORK)
12. Lastly, explain the usefulness of sinusoidal equations and their derivatives in the "real world." (Give at least two meaningful applications of sinusoidal equations.)

Lesson 8

Sequences and Triangular Differences

Our goal in the next few sections will be to devise a way to investigate and categorize certain data sets or ordered pairs in order to decide what kind of function models the data. To start, we will work with supplied data that can be modeled by the common functions taught at the secondary level. These functions are linear, quadratic, power, exponential, and logarithmic. Patterns will be identified in the domain set of a given function that lead to an identifiable pattern in the range of the function.

In order to use the function-identifying technique alluded to in the previous paragraph, we must first focus our attention on exploring some aspects of mathematical sequences generated by

$$f(n) = an^k,$$

where n is an element of the natural numbers, a is a real number, and k is an element of the positive integers.

Your first task, however, is to define what is meant by the term *sequence*.

Exploration 8.1: What is a Sequence?

1. Work in groups to settle upon and present a precise definition for the term *sequence*.
2. Also be prepared to report on where, in your mathematical careers so far, you have encountered and worked specifically with sequences and in what capacity.

You are now ready for the next Exploration in which you will investigate differences between consecutive terms of a defined sequence.

Exploration 8.2: Triangular Differences Involving Sequences of the Form

$$f(n) = an^k$$

1. Take a look at the sequence of square numbers listed in the first row of numbers below. The first several terms of the sequence are listed and, below that, the first differences of the two numbers immediately above and below that are the second differences of the differences (i.e., the second differences).

$$1 \ 4 \ 9 \ 16 \ 25 \ 36 \ 49 \ 64 \ 81 \ \ldots$$
$$3 \ 5 \ 7 \ 9 \ 11 \ 13 \ 15 \ 17 \ \ldots$$
$$2 \ 2 \ 2 \ 2 \ 2 \ 2 \ 2 \ \ldots$$

Notice that for the sequence n^2 the second differences are all 2. Now explore the differences for the sequence of cubes n^3. Do you notice any patterns? Make a conjecture as to what will happen with the fourth powers and the fifth powers. Next try to make a conjecture about the differences and the resulting constant for the sequence of kth powers of n^k where k is a positive integer. [This exploration can be done by hand, with a calculator, or in a spreadsheet program such as Excel.]

2. Is there a way of figuring out if there is a coefficient present such as cn^k where c is a constant for the sequence of kth powers cn^k?
3. Explain how the Triangular Differences process is related to the rate of change of your sequence function.
4. Explain how your findings for this activity are related to the differentiation rule for a certain type of continuous function from Calculus.

As a challenge activity, you might consider sequences generated by explicit equations containing multiple terms of the form cn^k. For example, try applying triangular differences for the sequences generated by, n, and $n^2 - 3n$. Do you notice any patterns? What about $2n^2 + 4n$ or $5n^2 + 2n - 5$? This exercise is most efficiently investigated using a computer spreadsheet program.

Lesson 9
Functions Defined by Patterns

Building upon what was learned in Exploration 8.2, we will now explore data sets consisting of ordered pairs. The purpose of this exercise is to detect patterns in the domain of a given data set that result in patterns in the range to identify the type of function that models the data. The type of patterns that we will look for will be arithmetic, geometric, or triangular difference patterns.

Exploration 9.1: Finding Function Patterns

The goal of this exercise is to try to use what you have learned from the sequence exploration that you have previously completed to find a pattern in each domain and related range of a given function in order to identify what kind of model may be present (i.e., exponential, linear, quadratic, …). At this point, we are not interested in the actual equation that models the data, only the kind of model present. Keep in mind that using triangular differences is not the only way to discern patterns. Look for ratios or multiplication patterns as well.

For each table of data:
1. Plot the data points and make a conjecture as to what kind of function is present.
2. Find a pattern in each domain and range for each function provided and try to understand how this pattern affects the shape of the graph that you have plotted (assuming that the graph is continuous).
3. Make a conjecture as to what kind of function can be used as a model for the given data.

Example 1

x	$f(x)$
2	4
4	9
6	14
8	19
10	24

Example 2

x	f(x)
1	15
3	5
5	19
7	57
9	119

Example 3

x	f(x)
1	15
3	135
5	1215
7	10935

It is a more complicated endeavor to find the pattern relationship in Example 4. Be aware that sometimes a pattern is revealed not by examining consecutive terms in a domain or range sequence; rather, consecutive terms can be considered as those terms, in order, that reveal a pattern. Consequently, sometimes ordered pairs can be "skipped" in order to discern a pattern relationship in the data. All ordered pairs will be points on the graph of the function. It is simply that the proper function pattern may not be revealed by examining each consecutive ordered pair.

Example 4

x	f(x)
3	135
6	1080
9	3645
12	8640

Example 5

x	$f(x)$
6	1
18	2
54	3
162	4

Functions Defined by Patterns—Verification

The patterns that you should have identified in Exploration 9.1 related to some of the more common functions studied in mathematics courses are

Addition-Addition: Linear
Constant Second Difference: Quadratic
Product-Product: Power
Addition-Product: Exponential
Product-Addition: Logarithmic

Of course, at this point your pattern identifications are simply conjectures. In the next Exploration, you are asked to verify these conjectures by working with general forms of the functions identified by pattern.

Exploration 9.2: Pattern Verification

In this exploration, you are going to work with general forms functions identified by pattern in order to discover why certain patterns in the domain of a given function lead to predictable behavior for the values produced in the range.

1. Addition-Addition: A Linear Function

   Statement: For a linear function f, adding a constant c to a given domain value results in adding a constant to the corresponding range value:

   To verify the statement above, find $f(x_2)$ in terms of $f(x_1)$ and fill in the statement below

$$\text{If } f(x) = mx + b \text{ and } x_2 = c + x_1, \text{ then } f(x_2) = \ldots\ldots$$

2. Add-Constant Second Difference: A Quadratic Function

   Statement: For a function f of the form $ax^2 + bx + c$ with domain values k units apart, then the second differences between consecutive $f(x)$ values are constant and equal to $2ak^2$.

   To verify the statement above, find the second differences involving $f(x)$ evaluated at

$$x_1, \ x_2 = (x_1 + k), \ x_3 = (x_1 + 2k)$$
.

---

3. Addition-Product: An Exponential Function

Statement: For an exponential function f, adding a constant c to a given domain value results in multiplying the corresponding range value by a constant:

To verify the statement above, find $f(x_2)$ in terms of $f(x_1)$ and fill in the statement below

$$\text{If } f(x) = ab^x \text{ and } x_2 = x_1 + c, \text{then } f(x_2) = \ldots..$$

4. Product-Product: A Power Function

Statement: For a power function f, multiplying a given domain value by a constant c results in multiplying the corresponding range value by a constant:

To verify the statement above, find $f(x_2)$ in terms of $f(x_1)$ and fill in the statement below

$$\text{If } f(x) = ax^k \text{ and } x_2 = cx_1, \text{then } f(x_2) = \ldots..$$

5. Product-Addition: A Logarithmic Function

Statement: For a logarithmic function f, multiplying a given domain value by a constant c results in adding a constant to the corresponding range value:

To verify the statement above, find $f(x_2)$ in terms of $f(x_1)$ and fill in the statement below

$$\text{Given } f(x) = a + b\log_n x, \text{ if } x_2 = cx_1, \text{ then } f(x_2) = \ldots.$$

Lesson 10
Using Functions Defined by Patterns In Application

Now it is time to explore how function patterns may be used in application. We will start with a fairly "well-behaved" situation in that it is not hard to discern patterns within the data provided. Our goal in a future Exploration will be to use function patterns and other learned techniques to investigate data for which finding a model is not so obvious.

Exploration 10.1: An Application of Functions Patterns

A slightly radioactive substance, called a "marker," is used to trace glucose metabolism in the heart. This substance has a half-life of about 3 hours. Suppose a dose of this marker were injected into a patient. Let $M(t)$ be the amount of the marker measured in *microcuries* (*mCi*) of that remains over time, t, in hours, as shown in the table.

t in hours	$M(t)$ in mCi
3	5
6	2.5
9	1.25

1. Determine the number of *mCi* that remain after 15 hours.
2. Use function pattern properties to make a conjecture as to the type of function that models the given data. What type of function models this pattern?
3. Why can't you use the pattern to find $M(22)$?
4. Find a particular equation for $M(t)$ (leave your equation exact) and verify that all of the $M(t)$ values in the given table satisfy the equation.
5. Use your equation to calculate the $M(22)$.
6. If another group presents a different equation that works for the given data, show that in fact, the different equation is an equivalent form of your conjectured equation. Otherwise, can you use concepts learned in a previous Exploration to find a model for the radioactive marker data?

EXERCISE: The "functions defined by patterns" process can also be used to create data sets specific to a desired type of function.

If a function f has values $f(5) = 12$ and $f(10) = 18$, use what you have learned about function patterns to find $f(20)$ if f is **a.** an exponential function; **b.** a linear function; and **c.** a power function.

UNIT TWO
REGRESSION AND MODELING

Lesson 11

Using Statistical Regression to Fit a Function to Bivariate Data

A regression method called *the method of least squares* will be used in this section to fit a *best-fit* linear function to data that are conjectured to be linear. A simple linear regression is accomplished by finding a line that minimizes the distance between actual data points and the predicted values $\hat{y}$ on the best-fit line. The difference between the predicted value at the ith data point $\hat{y}_i$ and the observed value y_i is called a *residual*, symbolized e_i, such that

$$e_i = y_i - \hat{y}_i.$$

The method of least squares used to find the predicted regression line employs an optimization technique from Calculus to minimize the sums of the squares of the deviations $y_i - \hat{y}_i$ such that an

estimated or fitted regression line of the form

$$\hat{y} = bx + a$$

can be produced.

An example of a scatter plot of data points and a best-fit regression line is displayed in Figure 11.1.

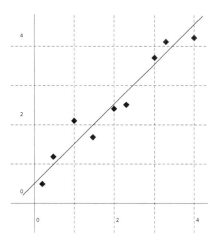

Figure 11.1. Scatter plot and regression "best-fit" line

One can use technology to avoid some of the tedious mathematics involved in producing the regression equation of the best-fit line. Graphing calculators, statistical software, or spreadsheet programs such as Microsoft Excel will calculate a linear regression equation to fit data.

Exploration 11.1: An Example of Linear Regression—Thunderstorms

This Exploration should be done as a full class exercise as a way for your instructor to teach you how to perform a linear regression investigation on relevant technology.

It is conjectured that in a lightning storm, the distance between one and the lightning is linearly related to the time interval between the flash and the bang. Consider d the distance to the storm in kilometers as a function of time t in seconds. Suppose, as an experiment, a friend travels along with the storm and reports the actual distance that the storm is from your house as you record the seconds between the flash and the bang.

1. Make a scatter plot of the data and use linear regression to write the particular equation for this direct-variation function.

t	2.9	6.1	14.9	28.9	37.2
d	1.0	2.0	5.0	10.0	12.0

2. Use your model to work backwards in order to calculate the times for the thunder sound to reach you from lightning bolts that are 1.5, 2.5, and 15 kilometers away. What do you call the processes of looking within and beyond your actual data?

CHALLENGE: What would your linear equation be with seconds and *miles* as your units?

Exploration 11.2: An Example of Linear Regression—Charles Law

This time perform this Exploration in groups or on your own to practice the use of relevant technology to perform the linear regression process.

Physicist Jacques Charles (1746–1823) discovered that the volume of a gas at a constant pressure is linearly dependent on the temperature of the gas. The table below illustrates this relationship. In the table hydrogen is held at a constant pressure of one atmosphere. The volume V is measured in liters and the temperature T is measured in degrees Celsius.

T	-40	-20	0	20	40	60	80
V	19.15	20.79	22.43	24.08	25.72	27.36	29.00

1. Consider V as a function of T and make a scatter plot of the data.
2. Use the table above and what you have learned about linear regression to find a model of the linear relationship.

Assuming that one started with 1 mole of gas at constant pressure, have you seen the values that you found for the T coefficient and the constant in the equation before? If not, you might consider doing some background research relating to Gas Laws.

3. Solve the equation that you have found for T to find

$$\lim_{V \to 0^+} T.$$

Have you seen the value that you found for the limit before? In what context?

In fact, through use of similar methods to that of the Method of Least Squares, technology can be used to find regression equations to fit linear, quadratic, power, exponential, logarithmic models, or other functions to supplied data. In this sense, theoretically, one could fit many different types of functions to a given set of data. So the question should be asked, "How, then, does one know which is the best type of model or function to fit to a sample of data points?"

Luckily, there are various "tools" available that help one decide which type of regression to use to fit a model or function to data. One tool that can be applied is the "functions defined by patterns" method explored in previous sections. Another tool available to us is the *correlation coefficient* associated with linear regression. The correlation coefficient r measures and describes a relationship or trend between two variables. Note that a correlation between two variables does not imply a causal relationship between the two variables. The relationship can, however, be characterized as having a positive correlation or a negative correlation, depending on whether the linear model associated with the data is increasing or decreasing, respectively. The correlation coefficient r can take on the values $-1 \le r \le 1$.

The value for r is obtained by computing the ratio of a measure of covariability between variables, divided by a measure of the product of variations within data associated with each of the two variables under investigation. The actual derivation of the formula for r is beyond the scope of this course. If the value of r is close to negative one, this indicates a strong linear correlation between variables from data with a negative trend. If r is close to positive one, a strong linear correlation between variables from data with a positive trend is indicated. A value of r close to zero indicates little linear correlation between the variables under investigation.

It must be noted that most graphing calculators and statistical and spreadsheet programs are programmed to display the value of the correlation coefficient r when one performs a statistical regression. However, other types of correlation between variables are also reported depending on the type of regression performed (i.e., linear versus other types of function regression). These are summarized here. Your instructor may choose to discuss these coefficient values at length with you.

r^2—The Coefficient of Determination

The value r^2, the *coefficient of determination*, measures the proportion of total variation in the values of Y that can be accounted for or explained by a linear relationship with the values of the random variable X.

R^2—The Coefficient of Multiple Determination

Both graphing calculators and software programs report an R^2 value for some non-linear regression models. This value is called the *coefficient of multiple determination*. It is analogous to the *coefficient of determination*, r^2, that applies strictly to linear situations. R^2 can be thought of as a multiple-linear least-squares fit. For example, if $Y = ax^2 + bx + c$, then Y can be considered *linear* in the coefficients a,

b, and c with $Y = aX_2 + bX_1 + c$ *where* $X_2 = (X_1)^2$. This same argument can be used for higher-degree polynomials.

This quantity represents the proportion of the total variation in the response Y that is explained by the fitted model. The closer to 1 the R^2 value is, the better the "fit" of the chosen regression curve.

Lastly, be aware that an r value (correlation coefficient) may be reported by graphing calculators and software programs for non-linear functions such as logarithmic, exponential, and power functions that can be made *linear by transformation*.

Lesson 12

Residual Plots and an Application

Another tool available to you that can help justify the choice of function used to fit data is the *residual plot*. Recall that a residual e_i is the difference between the predicted value at the *i*th data point $\hat{y}_i$ and the observed value y_i. For any type of regression chosen, one can plot the residual corresponding to each data point as a function of the explanatory or generating value of the bivariate data.

If the correct regression model has been chosen, one would expect the residual plot to consist of a scatter plot of values nicely distributed above and below the value $e = 0$ as pictured in Figure 12.1.

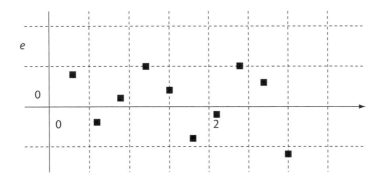

Figure 12.1. Residual Plot

On the other hand, if the regression model that you have chosen to fit your data is not truly the best-fit equation, the residual plot associated with the predicted and actual data will display a definite pattern as is depicted in Figure 12.2. Why do you suppose this is the case?

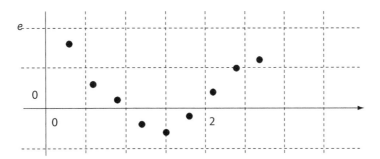

Figure 12.2. Residual Plot Displaying a Pattern

One must be aware of the fact that there is a limitation to using the residual plot as a tool for choosing which type of regression model is best to apply a data set. The limitation is that if one does not have enough data points in the set, then it is hard to tell if there is truly a pattern in the associated residual plot.

Exploration 12.1: An Application Activity Using Residuals

Average Weight and Length Measurements for a Common Rockfish Species

Age in years	Length in centimeters	Weight in grams
1	5.2	2
2	8.5	8
3	11.5	21
4	14.3	38
5	16.8	69
6	19.2	117
7	21.3	148
8	23.3	190
9	25.0	264
10	26.7	293
11	28.2	318
12	29.6	371
13	30.8	455
14	32.0	504
15	33.0	518
16	34.0	537
17	34.9	651
18	36.4	719
19	37.1	726
20	37.7	810

1. Our conjecture is that this is an exponential situation. Perform an exponential regression using WEIGHT in grams as a function of LENGTH in centimeters. Draw a graph of the scatter plot for the data that includes your regression curve.
2. Next, construct and draw a graph of the residual plot for the regression to confirm or reject our conjecture. Comment on the results.
3. Might there be a better model for these data? Test your answer to this question.

The next Exploration is an activity that will synthesize some of the various techniques used to find a function or model that best fits a data set.

Exploration 12.2: Kepler and Planets Exploration—Using "Real" Data.

The table below shows the periods of the orbits of each planet in years and the mean (half the sum of greatest and smallest distances or the semi-major axis) distance from the Sun in kilometers.

Name	Period P (yrs)	Millions of km from the sun (semi-major axis distance d)
Mercury	0.24	57.9
Venus	0.61	108.2
Earth	1	149.6
Mars	1.88	228.0
Jupiter	11.86	778.5
Saturn	29.46	1433.3
Uranus	84.01	2872.6
Nentune	164.79	4493.6

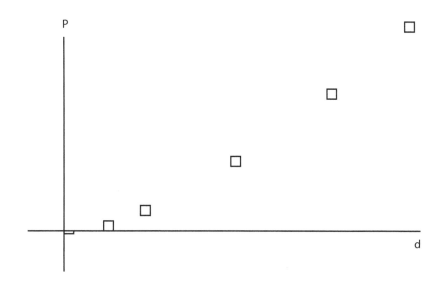

1. Use "function patterns" (remember these are real data) to try to decide what type of function these data represent considering *period, P,* as a function of mean *distance, d,* from the Sun (the semi-major axis of the planet's orbit).
2. Based on your answer to question 1, do a regression to find the equation for the function that fits this datum. In addition, comment on the r or R^2 value (whichever is appropriate based on your chosen model) associated with the equation.
3. Plot the scatter plot as shown above along with the regression curve to see how well the regression equation fits the data.
4. Now explore the data and regression equation using residuals. Does the residual plot support your choice for the type of regression that you chose? Are there problems with trying to interpret the residual plot for this exploration?
5. Kepler derived his three laws of planetary motion from analysis of data such as that in the table above. Research <u>Kepler's Third Law</u> in a reference text or in a physics text. Does your regression equation agree with that law?

Lesson 13
Terminal Speed Lab

The Terminal Speed Lab, like the previous activity, allows one to synthesize and review various topics and concepts explored so far in the course.

Falling Spheres and Terminal Speed Investigation

(The actual lab write-up should be done on separate paper.)

Purpose

To estimate the terminal speed of a falling sphere and find the distance the object falls in order to achieve terminal speed (note: we are not modeling the equation for the sphere's speed over time).

Required Equipment

stopwatch
long tape measure
lightweight ball (a Wiffle or Nerf golf ball would be good)

Background

Recall, from Calculus that the area under a graph of a speed (where speed is greater than or equal to zero) vs. time curve represents the distance traveled by the object. You will investigate the idea that the area under a graph of speed vs. time can be used to predict a property of the behavior of objects falling under the influence of gravity in the presence of air resistance. If there is no air friction, a falling Wiffle ball or Nerf ball falls at constant acceleration g so its change of speed is

$$V_F - V_o = gt$$

Where v_f = final speed
v_o = initial speed
g = acceleration due to gravity
t = time of fall

[NOTE: Although we can derive the needed formulas for the vector velocity, we will apply these derivations in terms of the non-vector *speed* of the ball. This is done simply to display our measurements as positive values for simplicity.]

A graph of speed vs. time of fall is shown at the top of the next page, where $v_o = 0$. The speed axis represents the speed v_f of a freely falling object (neglecting air resistance and wind direction, such that speed is measured as a positive entity) at the end of any time t. The area under the graph of the line is a triangle of base t and height v_f. Thus, the area equals $(1/2)v_f t$.

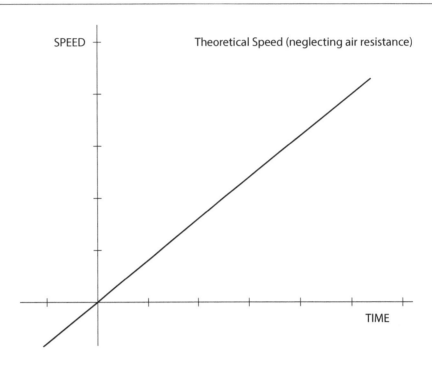

SPEED

Theoretical Speed (neglecting air resistance)

TIME

If you time a Wiffle or Nerf ball falling from rest a distance of 43.0 m in air, the fall takes 3.5 s. This is longer than the theoretical time of 2.96 s. Air friction is not negligible for most objects, including Wiffle balls. A graph of the actual speed vs. the time of fall looks like the curve below.

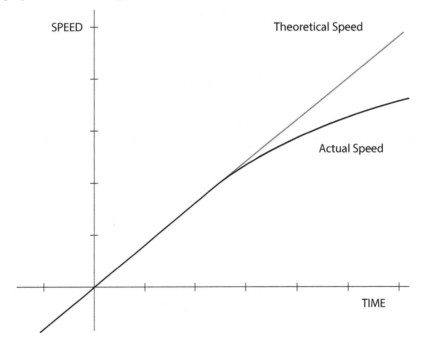

SPEED

Theoretical Speed

Actual Speed

TIME

Since air resistance reduces the acceleration of the object to below the theoretical value of 9.8 m/s^2, the falling speed is less than the theoretical speed. The difference is small at first, but grows as air resistance becomes greater and greater with the increasing speed. The graph of actual speed vs. time curve increases more slowly than the theoretical line.

Procedure

1. Using the Wiffle ball select five different sites from which to drop the object. Sites should range from 1 meter to 10 meters. You will clock the ball's time of fall to within 0.1 s. Consider various releasing techniques, and reaction times associated with the timer you use.

2. Practice your technique for dropping and timing to produce maximum consistency and avoid error in technique that could drastically affect your results.

3. Measure the height of each location and determine the falling times for your object at each site. Repeat three times per site and find the average time of fall for each location.

Some Equations

4. Using your measured value for the height, calculate the theoretical time of fall for your object at each location. Remember, this is the time it would take the object to reach the ground if there were no air resistance:

$$d = v_0 t + \frac{1}{2}gt^2 \quad where \quad v_0 = 0 \therefore \quad d = \frac{1}{2}gt^2$$

$$\Rightarrow t = \sqrt{\frac{2d}{g}}$$

5. Using your calculated theoretical time of fall from step 4, calculate the theoretical final velocity (speed, in our case) for an object falling without air resistance from each location you tested.

$$v_F - v_0 = gt \quad where \quad v_0 = 0 \therefore \quad v_F = gt$$

Graphing and Calculations

6. On graph paper, plot theoretical velocity vs. theoretical time. Draw a best-fit line between your data points. (See Useful Tables, pg. 4.)

7. Using your recorded actual times, calculate the actual final velocity of your object for each height.

$$d = \left(\frac{v_0 + v_F}{2}\right)t \quad where \quad v_0 = 0 \therefore \quad d = \left(\frac{v_F}{2}\right)t$$

$$\Rightarrow v_F = \frac{2d}{t}$$

8. On the same graph as in step 6, plot actual final speed vs. time. Starting from the origin, sketch your approximation for the actual speed vs. time curve. Your sketch should begin to level off after a certain amount of time. The limit this curve approaches is known as the terminal speed.

9. Using a computer spreadsheet program or graphing calculator, you will attempt to find an approximate value for the limit or the terminal speed. Use your graphing calculator to find a best-fit regression quadratic equation for your actual data. (Keep in mind that the actual curve is not really quadratic, but this method allows you to find a point where the derivative is equal to zero, which will be a good approximation for the time at which your object reached terminal velocity.)

10. Find the time at which the slope is equal to zero. This represents the maximum value for your curve. (This is an approximation for the time at which your object reached terminal velocity.) To find the point at which the slope is equal to zero, take the derivative of your best-fit regression equation that was generated from your calculator in step 9. Set the derivative equal to zero and solve for time. Once you have found this time, you can substitute this value into your original equation to find the terminal speed of your object.

11. By integrating your best-fit equation from time zero through time maximum from step 10, you can determine the fall distance at which your object will reach terminal speed. The distance to terminal speed is the area under the curve from time zero through time maximum on your graph.

Analysis
1. What can you say about objects whose speed vs. time curves are close to the theoretical speed vs. time line?
2. What does the area under your speed vs. time graph represent?
3. If you dropped a large leaf from the Empire State Building, what would its speed vs. time graph look like? How might it differ from that of a baseball?
4. The terminal speed of a falling object is the speed at which it stops accelerating. How could you tell by glancing at an actual speed vs. time graph if an object had reached its terminal speed?
5. Search references to find the actual equation that models how a small sphere falls in a medium and state this equation with a brief description. Also note how terminal speed fits into the equation.

USEFUL TABLES

Theoretical time	Theoretical speed	Actual Time	Actual speed
$t = \sqrt{\dfrac{2d}{g}}$	$v_F = gt$	t	$v_F = \dfrac{2d}{t}$
0	0	0	0

Lesson 14

Using Matrices to Find Models

Thus far in the course, we have had occasion to use some knowledge of trigonometric functions, solving of a few simple systems of equations, and statistical regression to find models to fit bivariate data. In this section, we will use matrices to solve certain systems of equations.

Suppose we have a 3 x 3 matrix A that represents a system of equations, along with a 3 x 1 solution matrix B. In this system, we are trying to solve for three unknown coefficients a, b, and c represented by the 3 x 1 matrix C. One might represent this situation with a matrix equation such as

$$[A][C]=[B]$$.

Keep in mind that our goal is to find the values of the coefficients of the C matrix. Recall that the way to accomplish this task is to multiply both sides of the equation by the inverse of the A matrix (provided that it exists) to yield

$$\left[A^{-1}\right][A][C]=\left[A^{-1}\right][B]$$,

where $\left[A^{-1}\right][A]$ is equal to the identity matrix, which simplifies to $[C]=\left[A^{-1}\right][B]$.

While it is clear that this process allows one to find the solution C matrix, one should ask, "Given an invertible system representing A matrix, how does one find the inverse of A?" The next example sheds light on the answer to this question.

EXAMPLE: How to Find the Inverse of A in a 3 × 3 Matrix

The goal is to find A^{-1} if

$$A = \begin{bmatrix} 1 & 0 & 1 \\ 2 & 1 & 3 \\ -1 & 1 & 1 \end{bmatrix}.$$

Begin by forming the following 3 × 6 augmented matrix. Perform matrix row operations to obtain the identity matrix on the left side, and perform the same operations on the right side of this matrix.

$$\begin{bmatrix} 1 & 0 & 1 & | & 1 & 0 & 0 \\ 2 & 1 & 3 & | & 0 & 1 & 0 \\ -1 & 1 & 1 & | & 0 & 0 & 1 \end{bmatrix}$$

$$\begin{matrix} R_2 - 2R_1 \to \\ R_3 + R_1 \to \end{matrix} \begin{bmatrix} 1 & 0 & 1 & | & 1 & 0 & 0 \\ 0 & 1 & 1 & | & -2 & 1 & 0 \\ 0 & 1 & 2 & | & 1 & 0 & 1 \end{bmatrix}$$

$$\begin{matrix} \\ \\ R_3 - R_2 \to \end{matrix} \begin{bmatrix} 1 & 0 & 1 & | & 1 & 0 & 0 \\ 0 & 1 & 1 & | & -2 & 1 & 0 \\ 0 & 0 & 1 & | & 3 & -1 & 1 \end{bmatrix}$$

$$\begin{matrix} R_1 - R_3 \to \\ R_2 - R_3 \to \\ \\ \end{matrix} \begin{bmatrix} 1 & 0 & 0 & | & -2 & 1 & -1 \\ 0 & 1 & 0 & | & -5 & 2 & -1 \\ 0 & 0 & 1 & | & 3 & -1 & 1 \end{bmatrix}$$

The right side is equal to A^{-1}. That is

$$A^{-1} = \begin{bmatrix} -2 & 1 & -1 \\ -5 & 2 & -1 \\ 3 & -1 & 1 \end{bmatrix}$$

It can be verified that $A^{-1}A = I_3 = AA^{-1}$.

Challenge: Based on this example, can you provide some general informal justification for why this process works?

NOTE:

If it is not possible to obtain the identity matrix on the left side of the augmented matrix by using matrix row operations, then A^{-1} does not exist.

Exploration 14.1: The Inverse of a Matrix

In this quick exploration, you are asked to find the inverse of matrix A where:

$$A = \begin{bmatrix} 0 & 1 & 2 \\ 1 & 0 & 3 \\ 4 & -3 & 8 \end{bmatrix}$$

How might you check your answer?

In the next exploration you will be asked to use what you have learned thus far in this section to find a model function or relation to fit data. This will be done in the context of working with two types of conics that have been previously discussed; the parabola and the circle.

Exploration 14.2: Using Matrices to Model Functions and Relations

- How many data points does one need to use matrices to find an equation for a parabola of the form:
$$y = ax^2 + bx + c$$
- Recall $[A]^{-1}[B] = [C]$, explain why this will work.
- How many points does one need to use matrices to find an equation for a circle of the form:
$$0 = x^2 + y^2 + Dx + Ey + F$$
- Is the above situation a function or a relation? Why?
- Find the equation for a circle containing points: (0, 0), (0, 5), (3, 3)
- Find the equation for a circle containing points: (-1, -3), (-2, 4), (2, 1)
- Find the equation for a parabola containing points: (0, 0), (20, 47), (30, 88)
- Find the equation for a parabola containing points: (0, 1.60), (10, 1.85), (20, 2.00)
- Lastly, put all equations found above into Standard Form

EXTENSION: Now consider
$$0 = Ax^2 + Cy^2 + Dx + Ey + F$$
This is the GENERAL Form of the equation for any of the Conic Sections that we have previously discussed (neglecting rotations).

- How would one know which conic is represented given an equation in this form?
- Could you put each of the conics represented in this form into the STANDARD Form for the specific conic?

Lesson 15

The Roller Coaster

The activity that follows serves to pull together various concepts that have been explored in the course so far. As you work through this activity, keep in mind that there are multiple approaches to achieving the desired outcome of the exploration.

Exploration 15.1: Building a Roller Coaster

An established company, the Thrill Ride Roller Coaster Company, has asked your group to help design a roller coaster track.

The Thrill Ride Company wants to build a roller coaster subject to a set of constraints. You are told that the company has several engineers who could design a blueprint and build the track if only they knew the functions whose graphs would define the desired curve of the track. According to a company spokesperson, her people can easily "fit" a curve to a set of points; however, the resulting curve does not necessarily satisfy constraints involving slopes, concavity, extrema values, smoothness, updateability, etc. Furthermore the company's engineers must first build a scaled-down test model, and thus they need the function description for the scale model. Hence your task is to define a **piecewise** function over the interval [0,15] whose graph satisfies the following constraints for the roller coaster track (each unit represents 10 meters).

1. The entrance onto the track is at the point (0, 10) and the exit is at (15, 0). There are just two local extrema, a minimum at (4, 2) and a maximum at (8, 8). (You do not have to consider designing the stairs leading to the entrance.)
2. The slope of the curve at the entrance and exit points must be zero in order to facilitate getting on and off the roller coaster car.
3. IMPORTANT: The curve must be "smooth," meaning that the piecewise function must be differentiable over its entire domain.
4. For continued customer interest, maintenance, and future customization YOU MUST build the roller coaster track out of pieces. Based on the given constraints, you must also decide how many functions that you will "sew together" to build the track.

The Thrill Ride Company is skeptical of work that has not been refined, stating that whenever safety, cost, or time is involved, they will not accept the plans of a single person. Therefore, you are to work in a group of three or four people.

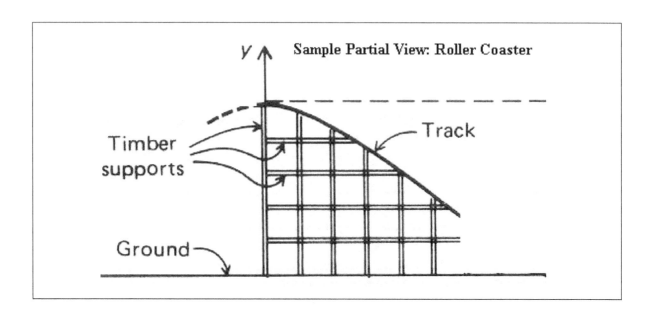

Sample Partial View: Roller Coaster

UNIT THREE
EXPLORING FUNCTIONS IN OTHER SYSTEMS

Lesson 16

A Non-Standard Exploration of the Rate of Change of Functions

W̲e will open this unit without preliminary explanation by having you complete the next Exploration involving a further look at functions and rate of change.

Exploration 16.1:

For this Exploration, we are concerned with the movement of an object along a path in the plane. We are assuming that the plane is a coordinate plane and the object starts at the point $(0, 0)$.

As the object moves along the path, each point on the path has two coordinates. The coordinates depend on the <u>distance traveled along the path</u>. Let us call this distance S. That is, S is the length of the path from the origin $(0, 0)$ to a point P on the path as depicted in Figure 1.

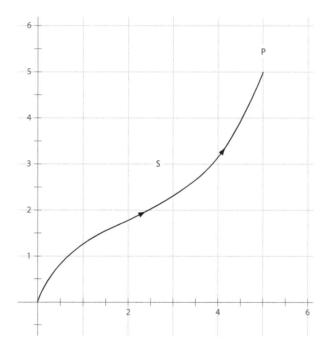

Figure 1.

Since the coordinates of P depend on the distance S we write $(x(s), y(s))$ for the coordinates of P.

We now wish to examine various types of paths <u>and</u> describe the behavior of the <u>functions</u> $x(s)$ and $y(s)$ as S increases.

Path 1 (A Polygonal Path)

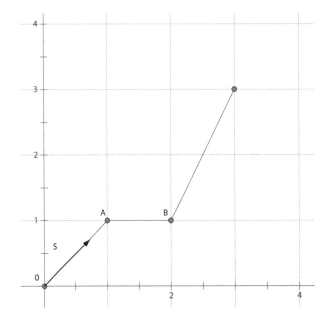

(a) Describe the behavior of $x(s)$ between the points O and A, A and B, and B and C.

(b) Describe the behavior of $y(s)$ between the points O and A, A and B, and B and C.

(c) Can you predict what the graphs of the $x(s)$ and $y(s)$ would look like for the entire polygonal path S between the points O and C?

(d) Is it possible to write explicit formulas describing the values of $x(s)$ and $y(s)$? For example, can we find $x(2)$ and $y(2)$? Also,

    (i) What value of s yields the coordinate point $(1.5, 1)$?

    (ii) What is $y(s)$ when $x(s) = 2.5$?

(e) Test your prediction from (c) by constructing graphs of $x(s)$ and $y(s)$. What is the domain of each of these functions, based on the supplied information?

Path 2 (A Parabolic Path)

Movement, in this case, is along a parabola with vertex at the origin. The simplest algebraic of such a parabola is $y = x^2$.

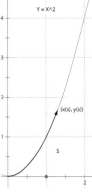

(a) Describe the behavior of $x(s)$, $y(s)$.

(b) Can we determine $x(2)$ and $y(2)$ from the given information?

(c) Can we construct graphs for $x(s)$, $y(s)$ (are they the same graph)? How might you accomplish this task; i.e., what materials are needed for you to do this?

Consider using a length of string and the scaled length of the provided ruler.

[NOTE: One centimeter on the supplied ruler and one unit on the "Parabolic Path" graph can be taken as the same. Please keep in mind that, due to paper copying irregularities, the provided ruler is likely not an accurate representation of a standard measurement.]

Exploration 16.2: A Parameterization of Movement in the Plane

Often it is useful to investigate a rule or relation between ordered pairs in the plane as they relate to yet a third variable or *parameter*. In this investigation, movement of an object in the plane is described by *parametric equations*. Using a non-standard approach, we will describe the path of an object as it moves around the circumference of a circle. The relationship between circumference arc length s traversed by the object and associated values we will call $w(s)$ and $h(s)$ will be described. Using the origin as our reference point, the value of $w(s)$ will represent the horizontal distance in the plane and $h(s)$ will represent the vertical distance in the plane used to locate the endpoint of the path of our object as it moves in a counter-clockwise direction along the circumference of a unit circle. The assumption will be that the chosen portion of the circumference is measured from the point $(1, 0)$.

To start, a circle of radius approximately one inch (for simplicity) centered at the origin of your coordinate system has been constructed in Figure 1. A thin piece of string of approximately 13 inches in length will be needed for this exploration. [NOTE: The scale on the supplied ruler and "one inch" unit circle can be taken as the same. Please keep in mind that, due to paper copying irregularities, the provided ruler is likely not an accurate representation of a standard inch measurement.]

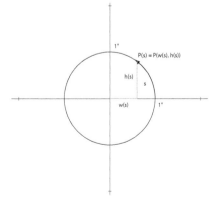

Figure 1. (Not drawn to scale)

Exploration procedure:

Place one end of a string at the point (1,0) and run a length of the string along the circumference of the provided circle. Take approximate measurements of your *s* lengths at rational multiples of 2π proceeding around the circle in a counter-clockwise fashion. Record readings in this manner starting from zero to twice around the circle. Measure your *s* string lengths using the provided ruler of Figure 2.

Figure 2. [Ruler: use inches to measure string]

Since you are approximating the various chosen *s* lengths, it might be helpful to note that 1/8 of an inch is equal to 0.125 inches and thus 1/16 of an inch is approximately equal to 0.063 inches.

In addition to recording the chosen circumference *s* lengths, one needs to also record the $w(s)$ and $h(s)$ distances associated with each *s* length as pictured in Figure 1. Recall that $w(s)$ represents the horizontal distance in the plane and $h(s)$ represents the vertical distance in the plane used to locate the endpoint of an arbitrarily chosen length *s* along the circumference of your circle. It will be helpful to create a table of values in the format of Table 1. Keep in mind that you must denote the positive or negative direction of your recorded $w(s)$ and $h(s)$ segment lengths depending on which quadrant you are working in when recording values. It is suggested that you take ten to twelve measurements as you progress around the circumference of the circle.

Table 1		
s	$w(s)$	$h(s)$
s_1	$w(s_1)$	$h(s_1)$
s_2	$w(s_2)$	$h(s_2)$
.	.	.
.	.	.
.	.	.
s_n	$w(s_n)$	$h(s_n)$

1. Describe any patterns you notice in the relationship between *s* and $w(s)$ and between *s* and $h(s)$ respectively.
2. Next, use the data recorded in Table 1 to construct two different graphs. The first graph should be a graph of $w(s)$ values as a function of the associated *s* values, and the second graph should be

of $h(s)$ values as a function of the associated s values. The graphs of Figures 2 and 3 are provided for your use. Be sure to label and scale each graph.

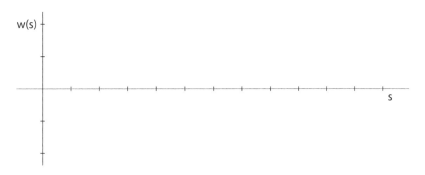

Figure 2.

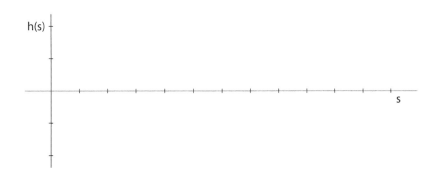

Figure 3.

3. Fully describe the pattern of each of the relationships now that you have graphed them.
4. Can you find the value for s such that $w(s)$ and $h(s)$ are of equal value?
5. Can you find the value for s such that $w(s)$ is half that of s? Also, what is the value of $h(s)$ at this point?

Points for Discussion:
(a) Choose four arbitrary ordered pairs $(w(s), h(s))$ from your list above and square each of $w(s)$ and $h(s)$. Then add the resulting values together. Take the average of your results simply to adjust for measurement errors. What does this average represent?
(b) We are using a circle of radius 1 unit centered at the origin. What is the effect of using a circle of different radius centered at another point?
(c) Given what you know circles and their properties, is there another way the parameter s could be named?
(d) Is it possible to estimate the distance traveled along the circle in terms of the length of the circumference of the circle?

Challenge:
(e) Perform the same exercise as you have just completed using an ellipse centered at the origin with major axis 2 units and minor axis 1 units. What observations can you make about these situations compared to the unit circle?

(f) Perform the same exercise as you have just completed using the right branch of the ellipse $\dfrac{x^2}{16} + \dfrac{y^2}{4} = 1$. What observations can you make about these situations compared to the unit circle?

(g) Perform the same exercise as you have just completed using the right branch of the hyperbola $x^2 - y^2 = 1$. What observations can you make about these situations compared to the unit circle?

Lesson 17

More Information Needed

Based upon your experience with the last section, you might have a better idea as to how to approach Exploration 17.1.

Exploration 17.1: Another Position-Time Relationship

At 1:00 p.m., a ship is 10 miles due east of port. At 2:00 p.m., it has sailed to a point that is 20 miles east and 50 miles north of the position at 1:00 p.m. Assume that the ship continues to sail in this manner as it did from 1:00 p.m. to 2 p.m.

A. Draw a picture or diagram to represent its voyage.
B. Place on your drawing the position of the ship at 3:00 p.m. and at 5:00 p.m.
C. How far north from the port will the ship be at 4:00 p.m.?
D. How far east from the port will the ship be at 4:00 p.m.?
E. Write a function that will give the ship's position at any given time.

As a result of completing Exploration 17.1, I hope that you realize that it is not to hard to create a function that describes the ship's *north position* as a function of its *east position*. The challenge occurs when one tries to introduce the additional information parameter of time into the situation. By now, you all should have realized that, in the last few sections, we are focusing our attention on the concept of *parametric equations*.

Previous to this Unit we have worked with equations such as $y = 4x^2 + 2x - 5$, where the variables x and y are related in a direct way. However, as a result of the Explorations of this Unit thus far, it should be clear that another way to describe a curve in the plane is to describe the x and y coordinates of a point on a curve in terms of a third *parameter* or variable. This third parameter is usually denoted as t. Thus, a *parallel set* of equations used to describe a curve can be given as:

$$\begin{cases} x = f(t) \\ y = g(t) \end{cases}$$

where f and g are functions of the parameter t. This parallel set is called the *parametric equations* of the curve. Note that parametric equations can be quite useful for modeling the interactions of vectors. We will encounter this application of parametrics in future sections.

Your instructor may wish to expand upon the examples this section and further relate some of these equations to trigonometric concepts that you have also studied in previous courses. Your discussion or consideration of these topics might also include techniques for switching between parametric and Rectangular representations of curves in the plane. In preparation for the next Exploration, a review of the *Parametric Chain Rule* from Calculus would also be in order.

Lesson 18

Applications Involving "More Information"

Exploration 18.1: Applications

A computer game company needs a clock to be integrated into a game scene. You have been hired to mathematically create a working clock with the shown radius (based on the second hand). The tip of the second hand's distances x cm and y cm from the left side of the shown box and the bottom of the box, respectively, depend on the number of seconds, t, since the second hand was pointing straight up. The second hand on the clock must turn with a period of 60 seconds and turn, of course, in the clockwise direction.

1. Create parametric equations for x and y in terms of t that satisfy the given conditions that the center of the clock is 25 cm from the left edge of the box and 26 cm above the bottom of the box with a radius of 17 cm.

2. At what rates are x and y each changing when $t = 10$ sec?

3. What is the slope of the circular path traced by the tip of the second hand when $t = 10$ sec?

4. Eliminate the parameter t and write the Rectangular equation of the circle that models the shape and location of the clock relative to the box.

Parametric equations can be used to model more advanced situations such as projectile motion.

5. Given

$$h(t) = h_0 + v_0 t - \frac{1}{2} g t^2$$

(h = height; v_0 = initial velocity; g = gravity; t = time; h_0 = initial height), use parametric equations to explore the change in distance horizontal x, or vertical y, or both over time t of a projectile fired from (0, 0) at an angle of 30^0 with an initial velocity of 45 m/sec.

6. Graph this situation to see if you, in fact, have the correct model.

Lesson 19

A Lab Involving Vectors

As was previously noted, parametric equations are quite useful when working with situations involving *vectors*. Recall that a vector is a quantity that has both magnitude and direction. As you work with the lab of this section, think about the connection between vectors and parametric representations.

Exploration 19.1: Vector Force Table Lab

Purpose

In this lab we explore Newton's 1st Law: $\sum \vec{F} = 0$ at equilibrium and Newton's 2nd Law: F = ma using vector forces on a force table. For the first half of the experiment, equilibrium for three vectors will be investigated by resolving each force into its vector components. In the second part of the lab, vector addition will be explored by finding a resultant between two added vectors and verifying your result using a third vector. As you perform the lab, strive to visualize the connection between vector components and parametric equations.

Method

A force table system is used under specified conditions to simulate the addition and subtraction of vectors. The vectors are created by hanging various gram masses from strings that are draped over pulleys and connected to a moveable ring near the center of the force table. The ring, when suspended off the table and centered, represents equilibrium in the system. For our purposes we will ignore friction at the pulleys.

Force Table System

Procedure & Analysis—Part I

1. Set up your vector table as shown above making sure that the middle ring is in equilibrium (i.e., the ring is not touching the center post on the vector table). Be sure to use various masses at each pulley to create your vectors at different angles.

2. For each of the vector string forces, the magnitude of the force can be found by using **F = ma** where **m** is the mass of the gram masses **measured in kilograms**, and **a** is the acceleration due to gravity: **9.8 m/s²**. Show your calculations for finding the magnitude of each of the three vector forces $\vec{v}_1, \vec{v}_2, and \; \vec{v}_3$.

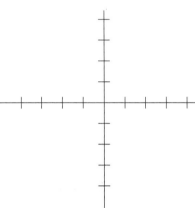

3. Using a protractor, draw your vectors *precisely* onto the coordinate system provided below. The angles of the vectors from the horizontal should be correctly shown and the magnitudes of the vectors should be proportionately represented.

4. Now resolve all three vectors into $\sum F_x \; and \; \sum F_y$. If you truly have equilibrium, what should each of these two sums equal? Show calculations to support your answer.

Procedure & Analysis—Part II

5. For this part of the lab you will use the same set of three vectors from Part I.

a) Now resolve **any two** of the vectors into $\sum F_x \; and \; \sum F_y$ to find the **resultant vector**. Show your calculations.

b) Once again, using a protractor, draw your vectors exactly onto the coordinate system provided below AND include your drawing of the resultant vector associated with the addition of the two vectors you chose to add in question a) above.

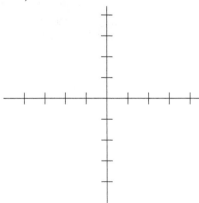

6. What should the relationship be between the resultant vector found in step 5 and the third vector in your system? Did you find the conjectured relationship to be true for your system? Support your answer.

Conclusion

7. Was there error present in your system? If so, what are some of the sources of error?

8. What conclusions are you able to make concerning your results keeping in mind the "purpose" of the lab from above?

Lesson 20

The Golf Shot

"The Golf Shot" activity is another Exploration application that will allow you to use and connect a few different concepts that have been covered thus far in the course.

Exploration 20.1: Tiger's Golf Shot

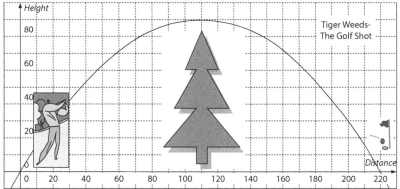

Earlier today, at the Golf Classic Open, Tiger Weeds hit a chip shot from the rough that just skimmed the top of a 90-foot pine tree and went right into the hole, 220 feet away, for an eagle on the 12th fairway.

Max. height: 90 feet   **Max. distance: 220 feet**

1. Use regression to find an equation for the path of the golf ball.
2. Now use matrices to find the equation for the golf ball's path.
3. Write the equation for the golf ball's path using the vertex form of the equation for a parabola.

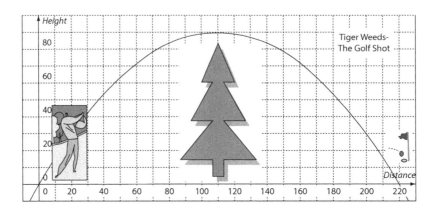

3. What is the angle at which the golf ball takes off?

4. In order to introduce the parameter of TIME t into this situation, first find the time it takes the ball to fall to the ground from its maximum height. Then find the ball's speed when it hits the ground.

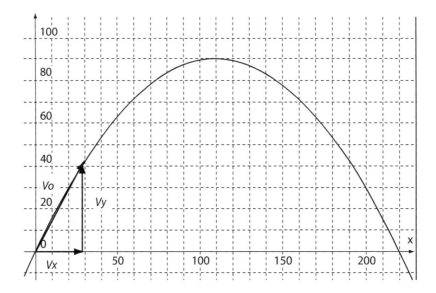

5. Last, use the diagram above and the information that you have found to write a parametric relation for the ball's path that incorporates time and relates this parameter to the distance and height that the ball travels.

Lesson 21

A Non-Standard Exploration of the Polar Coordinate System

It is assumed that you have had previous experience with the Polar Coordinate System. Your instructor may, however, wish to perform a quick review/refresher of the basics of the system at this point. The standard introduction to the Polar system usually involves a detailing of how one locates points in this system and then a jump to looking at classic polar relations involving trigonometric functions such as the Polar roses, cardioids, lemniscates, and circles. One then explores how to convert from Polar system to the Rectangular and back.

The Exploration of this section explores the Polar system in a completely different manner that does not depend upon conversions or trigonometric functions. Rather, this Exploration uses comparison as the key tool for understanding the Polar system in relation to the Rectangular system. This is accomplished by looking at "similar" functions in each system. Estimation and comparison (or commonality) arguments are mostly used to lead to understanding of the objectives of the exploration.

Keep in mind that the understanding of two basic concepts is needed to conduct the Exploration of this section. The first concept needed is a very basic understanding of vectors only in that one needs only to think of a vector as a *directed line segment*, which is a line segment to which a direction has been assigned. The second is an understanding of what a *radian* is. Recall that a radian is defined as an angle such that when its vertex is placed at the center of a circle, its sides intersect an arc whose length is equal to the radius. Thus there are 2π radians along the circumference of a given circle.

Exploration 20.1: Graphing Cartesian Functions in Polar Coordinates

Doppler radar used on television to report weather conditions; radar screens used by air traffic controllers to monitor aircraft traffic at an airport; flight plans filed by private aircraft to indicate paths taken as they move from one point to another; sonar positioning techniques employed in submarines; distances and compass bearings for directions used by campers in wilderness areas—these are but a few examples of the occurrence of vectors and polar coordinates in everyday life.

In this exploration we will use vectors to graph familiar equations in Cartesian coordinates and compare those to the equivalent graph in polar coordinates. We will use "(,)" for Cartesian coordinates and "$\langle \ , \ \rangle$" for polar coordinates. In Cartesian coordinates, a vector will represent the directed line segment from the point $(x, 0)$ to the point $(x, f(x))$ while in polar coordinates, a vector will represent the directed line segment from the pole $\langle 0,0 \rangle$ to the point $\langle f(\theta),\theta \rangle$. In the examples and exercises the domain of the functions will be limited to the set of nonnegative real numbers. We will explore both linear and quadratic expressions.

Exploration:
Linear Expressions

1. (Example for your consideration) We begin with a constant function $y = c$, where $c > 0$.

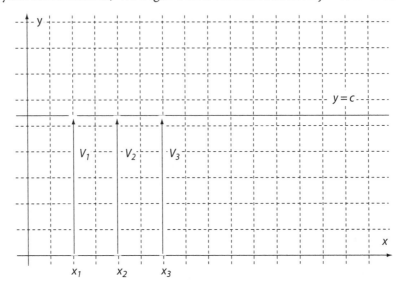

Figure 1a.

In this example, the vectors in Cartesian coordinates easily translate to vectors of fixed length bound at the origin with the tip of the vector lying on a circle of radius c.

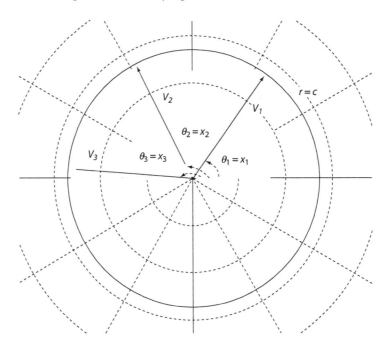

Figure 1b.

2. One should have little difficulty in graphing $y = x$, and can use this to interpret what should take place with the graph of $r = \theta$ for $\theta \geq 0$. Use the "vector approach" of the previous example as a guide to do this.

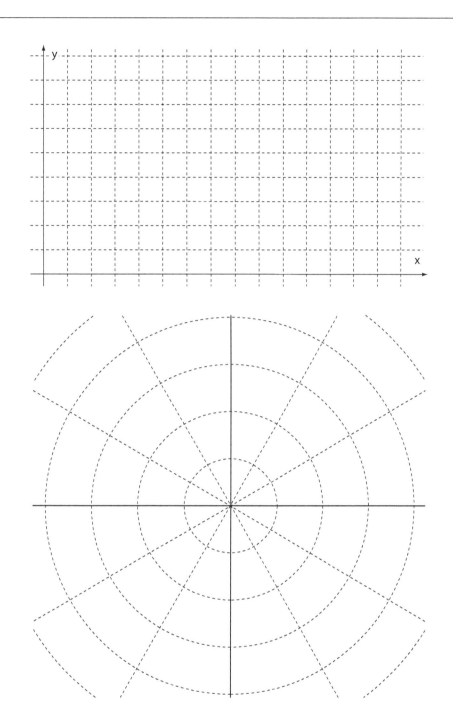

Quadratic Expressions:

We next consider polar quadratic functions of the form: $r = (\theta - a)(\theta - b)$, where $0 < a < b$; $r = (\theta - a)$, where $a > 0$; and $r = \theta^2 + a\theta + b$, where $r(0) \neq 0$ for all θ.

3. Use the same vector approach to graph $y = x - 3x + 2 = (x - 1)(x - 2)$ and the corresponding polar graph $r = (\theta - 1)(\theta - 2)$ on the grids provided below (you may have to adjust your scale on each axis). The vertex of the parabolic graph is at $\left(\frac{3}{2}, -\frac{1}{4}\right)$ with axis of symmetry at $x = \frac{3}{2}$. For

the polar graph, consider three rays corresponding to the values of $\theta = 1$ rad, $\theta = \dfrac{3}{2}$ rad, and $\theta = 2$ rad.

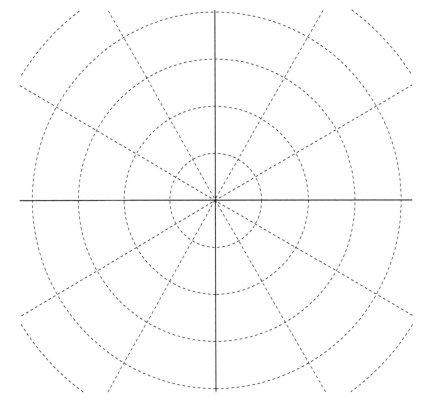

4. Consider a quadratic that has only one positive real root, $y = x^2 - 4x + 4 = (x - 2)^2$ and the corresponding polar curve, $r = \theta^2 - 4\theta + 4 = (\theta - 2)$. When constructing the polar graph there is one important ray to consider, the ray $\theta = 2$ rad. Use the same vector approach to explore the connection between these graphs in the two systems.

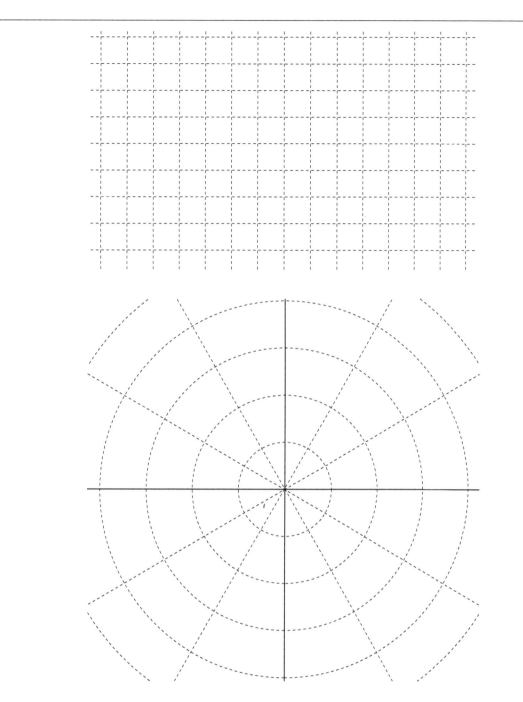

5. Next consider the quadratic, $y = x^2 - 4x + 8$ which has no real roots and is positive for all values of x. Perform the same systems exploration using the Cartesian and polar grids below.

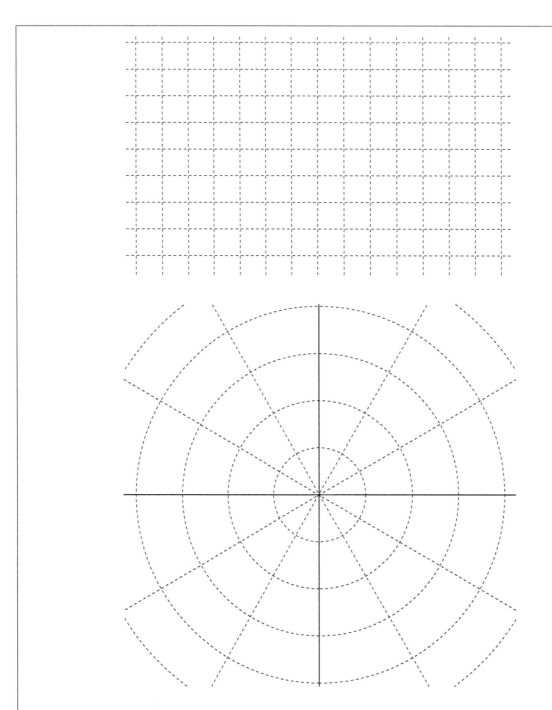

Extension:

6. Use what you learned previously about Rectangular-Polar graphing connections and the Cartesian graph provided in Figure 2 to create a "polar version" of the graph for $0 \le \theta \le 6$.

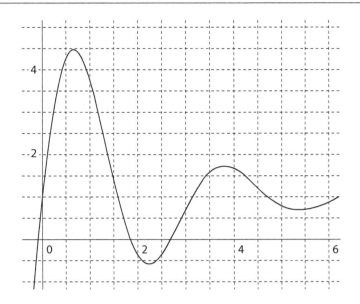

Figure 2.

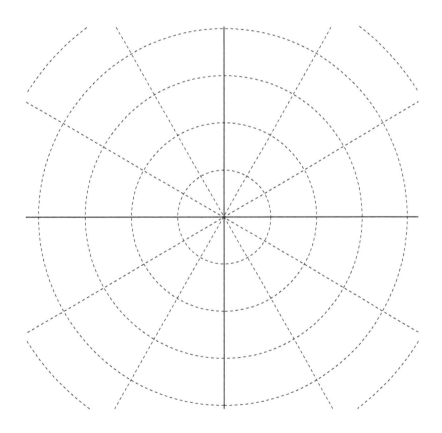

Historical Note:

While ancient Greek mathematicians such as Archimedes made references to functions of chord length that depended upon angles measured, it was a Persian geographer, Abu Rayhan Biruni, circa 1000, who is credited with developing an early foundation for a polar coordinate system. The polar coordinate system as known and used today, however, is credited as having been developed by Issac Newton circa 1671, and further refined and used by Jacob Bernoulli circa 1691.

Lesson 22
The Geometry of Complex Numbers

As a warm-up for the Exploration of this section, your instructor should first conduct a short review of the field properties of the complex number system and have students try to come up with a way to visualize addition and subtraction of complex numbers in the coordinate plane. It will also be helpful to understand what is meant by the *modulus* of a complex number. The *modulus* or absolute value of $z = x + yi$ is represented by the symbol $|z|$ and is defined as

$$|z| = \sqrt{x^2 + y^2}.$$

Exploration 22.1: The Geometry of Complex Numbers

1. Now that you have reviewed the geometry of complex number addition and subtraction along with the *modulus*, $|a + bi|$, of a complex number; try to provide **geometric evidence** of the triangle inequality for complex numbers:

Theorem A: For any complex numbers w and z, $|w + z| \leq |w| + |z|$.

2. Use **Theorem A,** above, to algebraically prove an extension of the triangle inequality for complex numbers, namely

Theorem B: For any complex numbers w and z, $|w - z| \geq |w| - |z|$.

3. For any $z = a + bi$, we can define $\overline{z} = a - bi$ as the *conjugate* of z.

   a. If w and z are complex numbers, show that $\overline{w + z} = \overline{w} + \overline{z}$

   b. If w and z are complex numbers, show that $\overline{w \cdot z} = \overline{w} \cdot \overline{z}$

   c. Show that $|z| = |-z| = |\overline{z}|$ for all complex numbers z.

   d. Show that $|wz| = |w| \cdot |z|$ for all complex numbers w and z.

   e. Write $z \cdot \overline{z}$ in terms of the modulus of z.

Definition: If w and z are complex numbers ($z \neq 0$), then $\dfrac{w}{z}$ is the complex number u such that

$w = z \cdot u$.

      f. Find $\dfrac{w}{z}$. [Leave your answer in terms of z and w and verify the result.]

      g. Multiply your answer above by z. Comment on why this makes sense with regard to the concept of "division."

      h. Prove that $\left| \dfrac{w}{z} \right| = \dfrac{|w|}{|z|}$, *and that* $\left| \overline{\dfrac{w}{z}} \right| = \dfrac{\overline{|w|}}{|z|}$

Lesson 23

Complex Numbers in Polar Form and Euler Numbers

Thus far in this course, among other topics, you have explored the Polar Coordinate System and complex number properties. In this section, we will combine concepts learned about each of these topics in order to explore complex numbers in polar form.

A complex number $z = x + yi$ can also be written in *polar form*

$$z = r(\cos\theta + i\sin\theta)$$

where r is equal to $|z|$ and θ is defined as the *argument* of z since

$$x = r\cos\theta \quad \text{and} \quad y = r\sin\theta.$$

The argument θ is found by using the fact that

$$\tan\theta = \frac{y}{x},$$

where the quadrant containing z must be considered when finding θ.

Also note that $\cos\theta + i\sin\theta$ can be written as $e^{i\theta}$ by using *Euler's Formula*

$$e^{i\theta} = \cos\theta + i\sin\theta$$

Therefore, we have

$$z = x + yi = re^{i\theta}, \tag{23.1}$$

where the latter form in equation (23.1) is called the *exponential form* or *Euler Number*.

Using the information discussed to this point in the section, one is now ready to attempt Exploration 23.1.

*As a side note notice that, if one works with the power series expansions of e^x, $\cos x$, and $\sin x$,

Euler's Formula can be found by:

Start with

$$e^x = 1 + x + \frac{x^2}{2!} + \frac{x^3}{3!} + \frac{x^4}{4!} + \ldots \qquad = \sum_{k=0}^{\infty} \frac{x^k}{k!}$$ (1)

$$\sin x = x - \frac{x^3}{3!} + \frac{x^5}{5!} - \frac{x^7}{7!} + \frac{x^9}{9!} - \ldots \qquad = \sum_{k=0}^{\infty} \frac{(-1)^k x^{2k+1}}{(2k+1)!}$$

$$\cos x = 1 - \frac{x^2}{2!} + \frac{x^4}{4!} - \frac{x^6}{6!} + \frac{x^8}{8!} - \ldots \qquad = \sum_{k=0}^{\infty} \frac{(-1)^k x^{2k}}{(2k)!}$$

Now replace x in (1) with ix where i is the imaginary number. Thus

$$e^{ix} = 1 + ix + \frac{(ix)^2}{2!} + \frac{(ix)^3}{3!} + \frac{(ix)^4}{4!} + \frac{(ix)^5}{5!} + \frac{(ix)^6}{6!} + \frac{(ix)^7}{7!} + \frac{(ix)^8}{8!} + \ldots$$

$$= 1 + ix - \frac{x^2}{2!} - \frac{ix^3}{3!} + \frac{x^4}{4!} + \frac{ix^5}{5!} - \frac{x^6}{6!} - \frac{ix^7}{7!} + \frac{x^8}{8!} + \ldots$$

$$= \left(1 - \frac{x^2}{2!} + \frac{x^4}{4!} - \frac{x^6}{6!} + \frac{x^8}{8!} - \ldots \right) + i\left(x - \frac{x^3}{3!} + \frac{x^5}{5!} - \frac{x^7}{7!} + \ldots \right).$$

Therefore

$$e^{ix} = \cos x + i \sin x.$$

Exploration 23.1: Complex Numbers in Polar Form

A. Given two complex numbers in polar form

$$z_1 = r_1(\cos\theta_1 + i\sin\theta_1) \ \text{and} \ z_2 = r_2(\cos\theta_2 + i\sin\theta_2),$$

1. Show that $z_1 z_2 = r_1 r_2[\cos(\theta_1 + \theta_2) + i\sin(\theta_1 + \theta_2)]$.

2. Show that $\dfrac{z_1}{z_2} = \dfrac{r_1}{r_2}[\cos(\theta_1 - \theta_2) + i\sin(\theta_1 - \theta_2)]$.

Use the result from question 1 to present an argument supporting DeMoivre's Theorem, which states that if z is a complex number in polar form, then for any positive integer n

$$z^n = r^n(\cos n\theta + i \sin n\theta)$$.

B. Corollary to DeMoivre's Theorem: For any $z = r \, cis\theta$ and n any positive integer, the n distinct nth roots of z are given by $\sqrt[n]{r} \, cis\dfrac{\theta + 2\pi k}{n}$ for $k = 0, 1, 2, 3, \ldots, n - 1$.

1. Use the corollary above to find the five fifth roots of 1. Then graph these roots in the complex plane.

2. Use the theorem above to factor $p(x) = x^3 - 10$, into linear factors with complex coefficients.

EXTENSION: Use the previously discussed Euler Number $e^{i\theta} = \cos\theta + i\sin\theta$ to derive some common trigonometric identities.

Write $e^{i\theta_1} \cdot e^{i\theta_2}$ two different ways in order to derive the *sine* and *cosine* angle addition identities.

Now try to derive the *sine* and *cosine* angle subtraction identities by writing $\dfrac{e^{i\theta_1}}{e^{i\theta_2}}$ two different ways.

Conclusion

It is the authors' sincere hope that the Explorations contained within this text have enticed you to think deeply about some of the mathematics you've encountered previously, about new ideas presented, and about the connections between the two. It is also our hope that the inquiry-based "open-forum" methodology used to deliver this course has broadened your perspective and encouraged you to consider using inquiry-based teaching methods in your classrooms when you embark upon your professional careers as secondary mathematics instructors.

Best wishes,
The Authors

Index of Selected Items

CPSIA information can be obtained
at www.ICGtesting.com
Printed in the USA
LVHW060922290121
677767LV00006B/82